An Introduction to the Serama Bantam

The author as a 6 year old holding her favourite hen 'Snowy'.

This book is dedicated to my Mum, Dad and sister Claire for their encouragement and support over the years. Also a massive thank you to my brother Chris Jnr, as without him the Serama forum would not be what it is today.

Carrie with her brother and sister proving you are never too young to start enjoying poultry.

An Introduction to The SERAMA BANTAM

Carrie Wright

Seeright Publishing

First published 2009

Copyright © 2009 by Carrie Wright

Published by

Seeright Publishing
54 Glen Road
Lincs
NG33 4RJ
info@seramaforum.com

ISBN 978-0-9562255-0-4

No part of this book may be reproduced or transmitted in any way or by any means, electronic or mechanical, including photocopy, recording, or any information storage and retrieval system, without permission in writing from the publishers. All rights reserved.

Printed by

BJ's Print & Design
Stamford
Lincs
PE9 1XP

Contents

Introduction	2
History of the Serama Bantam	3
Housing Requirements	5
Choosing Your Serama Bantam	12
Feeding	15
Breeding, Incubation and Rearing	20
Exhibiting Your Serama	35
The Standard for the UK Serama Bantam	42
Problems and Ailments	45
When? What? How? Why?	51
References	54

Flock of Silkied and Straight feathered Serama.

Introduction

This book has been written for everyone who wishes to start keeping Malaysian Serama bantams. Much about the breed is still shrouded in mystery so the author's intention is to provide an insight into owning these charming little birds.

Serama bantams differ to ordinary chickens due to their small size, character and spirit – perfect for indoor living – in fact, in Malaysia Serama outnumber dogs as the preferred house pet! You will learn why in this book.

One of the most commonly asked questions concerns their housing requirements. Having originated from tropical Malaysia, Serama do require relatively warm conditions in order to thrive. They are a fairly delicate breed; however, with the right care, they cope well with our climate. Advice and recommendations are provided in the following chapters.

It is important to buy your birds from a reputable breeder. This will ensure they are full of vitality and have good, strong lines, providing an excellent starting base for breeding further quality specimens.

Breeding and rearing chicks is one of the most rewarding aspects of keeping any poultry, but when that task is complicated by tiny, vulnerable eggs, you are likely to become even more determined to succeed. Of course, the most reliable and successful brooder is the hen itself. Serama bantam hens do make great mothers although they cannot cover a large number of eggs.

The overall vigour and condition of the bird is enhanced by feeding techniques. There are a large variety of feeds available, but not all are suitable for the Serama bantam due to the size of the pellets and ingredients used. Some of the growers' pellets, for example, contain growth promoters that are more suited for large fowl and can be disastrous for bantams.

The compact size of the Serama is what differentiates them from any other bantam. Each bird must conform as closely as possible to the **UK Serama Bantam Standard of Excellence set by the Poultry Club of Great Britain.** The improvement and development of the breed is still ongoing but the standard assists the breeder in creating the 'perfect' bird. It also provides a template for the ultimate bird, allowing the judge to award points for each characteristic. The Serama is yet to breed true, and so variations in the offspring are very likely.

As with all poultry, problems and illnesses can occur – some more serious than others. It is vitally important to recognise the signs early, so that relevant treatment and care can be sought.

History of the Serama Bantam

The ancestor of all domestic chickens is a sub-species of the Red Jungle Fowl of Asia, *Gallus gallus gallus*. The first recorded domestication of chickens in Asia goes back about 8,000 years ago when they were mostly bred for cock fighting. From Asia they eventually reached Britain, where it is believed that the Phoenicians brought the fowl for Julius Caesar, who arrived in 55 BC. These too were used solely for fighting and it was not until later years that chickens were developed for their egg-laying abilities.

Centuries of selective breeding have created the many different varieties of chickens that we recognise today.

The Malaysian Serama Bantam is no exception, even though its complete ancestry is unknown, it is believed the breed dates as far back as the 17th Century but there is no solid evidence to support this.

Red Jungle Fowl

The breed's development is due to the determination and perseverance of Wee Yean Een whose early fascination with chickens encouraged him to pursue his dream. In 1971, Mr Een acquired some chickens similar to the Modern Game Bantam called *Ayam Kapans*. He wanted to enhance the bantam with a compact bone and body structure and so introduced the Silkie Bantam into his breeding plans, which surprisingly produced, smooth-feathered offspring. However, the Silkie's own characteristics, such as five toes and feathered legs, were undesirable so he tried to breed these out. Nevertheless, some of the Silkie Bantam traits often crop up due to throwbacks (showing characteristics of an earlier type) and even though five toes and leg feathering are not desirable, work is being carried out to improve the Silkied Serama.

Wee Yean Een then wanted to improve the tail conformation on his birds so he introduced the Japanese Bantam or *Ayam Jepun* into his breeding programme to create an erect, 90-degree tail that is prominent in the Japanese. Mr Een was hoping to achieve a regal, full-breasted bird with a compact body, combined with an erect tail and a

This Silkied Serama cockerel exhibits a good high tail, short back and excellent wing length.

vertical, low wing carriage. The programme was successful; he even managed to produce a smaller bird, which he favoured and developed further.

It is believed that the Thai Miniature Hen was also incorporated into the breeding plan because of its influence on size. There is not enough evidence to support this claim, even though the Serama and Thai Miniature share some similar genes that have resulted in the small size.

By 1988, these prototypes were beginning to improve and were now weighing less than 500 grams. The bird's weight was an important factor in the development of the breed and as its compact size was a vital characteristic it was decided that the Serama would be classed according to its weight.
These weights were divided into three classes – A, B and C, with Class A Serama being the smallest and progressing in size to Class C.

Now that Wee Yean Een had begun producing birds that were to his liking, he decided to name them, Serama Bantams. The name was derived from his childhood love of shadow puppet plays where a mythical character, Raja Sri Rama, was famous for his beauty and majestic attributes; ideal for the Serama Bantam. Keen to promote his newly named breed, Mr Een decided to sell off his surplus stock which provided him with a much-needed financial boost.

His venture was a great success and in 1990 the birds had become so well-known that the first, Serama Bantam Show took place in the district of Bukit Batu Pahat in Malaysia. Even though the show was combined with a state government-organised event, featuring songbirds and fighting game, it was a huge achievement for Wee Yean Een and it seemed only right that he should be the judge.
This very first show was the making of the Serama Bantam and its popularity grew so much that several shows would take place in a single week and it was not long before their appeal had spread to Thailand and Singapore. Their beauty and small size make them ideal as house pets and they live quite happily indoors alongside humans – which is why, amazingly, Serama now outnumber dogs and cats in Malaysian homes.

Two Serama hens.

The Serama Bantam is constantly being improved; perfecting size, temperament, posture and type. All of which are imperative to this enchanting and unique bird.

Housing Requirements

The accommodation is one of the most important aspects for any poultry keeping. There are many manufacturers that offer a vast range of poultry houses; from the smaller coop, suitable for a trio of bantams, right through to the larger house used for commercial poultry. Whatever the size, they all incorporate similar properties that provide a healthy living space.

Housing Indoors and Heating Requirements

Due to their small size, a trio of Serama bantams requires very little space and can be comfortably housed in a 24-inch by 18-inch cage. Many Serama owners house their birds in modified rabbit hutches or cages that can fit in the family home. This form of housing has its benefits and drawbacks.

A large rabbit cage - shown here housing Serama growers and quail.

The main advantage of keeping them in an indoor rabbit cage, is that they will become more confident and used to different sounds, consequently if you were to show the birds, they would not be fazed by the hustle and bustle of the event. Another benefit is that as the Serama would be in close proximity within the family home; they are more likely to be handled creating a very tame and trustworthy pet. It is even possible to house train a Serama so that it will learn to defecate in a specified area. This of course takes many months of patience and training and is achieved by returning your bird to the area (usually a litter tray) after it has defecated - eventually the bird will associate the litter tray with toileting. Rabbit cages are generally not very spacious, even for Serama bantams, so allowing them to venture out of their cage daily will enable them to stretch their legs and prevent boredom.

A small wooden house - large enough to accommodate a trio of Serama.

One thing that must be highlighted is that Serama housing should not be exposed to draughts, and this needs to be taken into consideration when using an open wire cage, as it will offer little protection unless combined with suitable additional coverage.

Cleaning this form of housing is very easy and it can also be washed and sanitised. One problem that can arise is, as with all poultry, Serama love to scratch and forage. Therefore if you are using a bedding, such as shavings, it is quite likely it will be scattered outside the cage, and this may not suit the house proud!

Some breeders cope well with using newspaper as a base but this does need to be changed daily and any wet or damp areas should be removed to prevent them from being consumed. Another common floor covering is fine sand. The sand must be cleaned and can be bought relatively cheaply from most pet stores. Once soiled, the sand can be sieved to remove any droppings or food from the cage and topped up with fresh sand as required. It is recommend refreshing all the sand after about 10 days, as harmful bacteria can breed and multiply if not discarded. You will find that using sand as bedding enables your Serama to bath regularly. This not only rids them of parasites but also conditions their plumage.

Another form of housing similar to the above, is a rabbit hutch. Again the same advantages apply but they can be smaller in size than the indoor cages. A two-tier rabbit hutch provides extra freedom but it is vital that the birds are allowed to exercise outside of the hutch. A thin layer of wood shavings can be scattered inside and this makes for easy cleaning. Adding a perch in the enclosure area turns this into an excellent roosting box or an ideal nest box.

A plastic or wooden base allows for

trouble-free cleaning and disinfecting. Often rabbit hutches have a hinged roof or removable front doors. Again, this all helps towards easy maintenance of the house. When positioning the rabbit hutch ensure that it is in a secure, sheltered and draught-free area indoors or weather permitting, outdoors.

A simple layout showing grass pens with small wooden coop – ideal for summer months.

In the warmer summer months, you may prefer to house your Serama in a movable coop and run. A wooden coop will provide enough protection against the cooler nights; therefore a heat source would not be necessary.

It is important to consider the coop design and structure, as there are so many on the market it is easy to make the wrong decision.

The large majority of coops are made from timber, as it is strong, readily accessible and relatively cheap to buy. A drawback with timber is that, if left untreated, it will deteriorate over time. However, various preservatives and treatments that can be used to increase the structure's longevity. The most popular method is to apply preservatives by means of controlled impregnation processes using vacuum cycles. This tanalised wood provides the most effective long-term protection against the elements and is non-toxic. Other techniques involve applying treatments by brushing, spraying or immersing the wood in preservatives. This however, only results in partial penetration and therefore gives limited protection. Another factor to bear in mind is whether the coop can be easily moved. Even a small wooden coop can be very heavy, so it is important that it has carrying handles, wheels or skids, and especially so if there is a connecting run, as this will need to be moved to fresh ground every few days.

Along with the timber, check the *roof*. Some poultry manufactures still use roof felt to cover the top of the house because it is easier to work with. The main disadvantage with felt is that it provides an excellent environment for mites and, in particular, red mites. Once these have taken up residence, the felt has to be removed and the whole coop treated, which can be a very time-consuming task.

A more popular choice is Onduline roofing sheets. These are tough, flexible and yet lightweight and manageable. Its corrugated surface allows ample light and ventilation to circulate the coop and there are no hidden areas to harbour mites. The only disadvantage with this form of roofing is that rodents can enter the coop via the ridged opening. Therefore it is wise to mesh over the gaps to prevent entry.

Once the outside of the coop has been assessed, it is important to study the inside. Serama bantams have very small feet, therefore the width of the perch needs to be taken into consideration. The best width is approximately 3 cms with slightly bevelled sides to suit the birds' grasp. Depending on the size of the coop you might be able to have more than one perch and if they are removable it will help when it comes to cleaning. The perch arrangement is better if they are positioned higher than the nesting area as this discourages the bantams from roosting and fouling the nest box. Serama bantams tend to only use one communal nest box and this is sufficient for a trio of birds. It should be lined with cleaned, chopped straw or with untreated wood shavings. The latter is preferable as straw can harbour mites and dampness can encourage mould growth. Never use hay as nest material as harmful spores can be expelled that are responsible for the chronic respiratory problem, Farmer's Lung.

The base of the coop may incorporate a *droppings board*. This catches the droppings, and because it slides out, cleaning is much easier and it also simplifies disinfecting and sterilising. If you decide to buy a coop without a droppings board, then a plastic sheet secured to the base can serve equally well. Again the best floor covering is wood shavings, although some owners like to use newspaper but this is less manageable and soon spoils.

It's vital that the coop is easy to clean. A large, wide door is best, but a hinged roof can also suffice, as long the coop is not too deep. There is nothing worse than having to cram your arms into a coop through a tiny door!

The *pop-hole entrance* that leads to the run needs to be operated from outside of the run. This can be achieved by attaching a length of wire or twine to the pop-hole door which can then be hooked back. Another process that is commonly used by manufacturers is to have a piece of wood attached to the pop-hole door that slides open when pushed. This is more pleasing to the eye but will add to the overall cost of the coop.

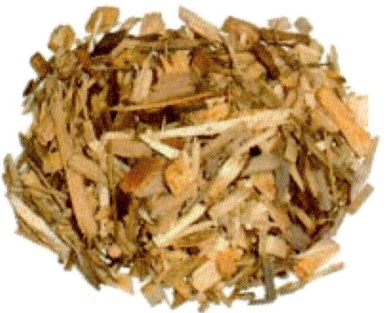

The *run* itself will provide a great source of enrichment for your Serama, particularly as the vegetation it incorporates will add necessary vitamins and minerals to their diet. Even the tiny claws of your Serama soon obliterate any sign of grass so the run must be moved every couple of days. This will also prevent any build up of droppings and therefore minimise the risk of disease.

If you would prefer not to rotate the run, then a thick layer of wood chips would provide an ideal base. These allow water to drain through, thus keeping the top relatively dry, and they are easily raked and maintained. It will be necessary to refresh with new chippings every 3 to 4 months depending on wear. Pea gravel also makes for an excellent floor covering, as it is more hygienic than wood chips and can be disinfected, plus it will not rot down into a mush.

Washed greens hung up in the enclosure are appreciated, but change them regularly as they soon wilt and become unappetising. Always be wary of offering your Serama wild plants, such as dandelion leaves, as these may well have been sprayed with a chemical or could have come into contact with wild birds that could be disease carriers.

The sides of the enclosure are generally covered in wire mesh. The mesh size differs from product to product, but smaller is better (approx. 2.5cms), as this will prohibit wild birds from entering the run. If rain if forecast, be prepared to sheet over the run to shelter your birds from the wet. This will also help to prevent the run from becoming waterlogged.

You might like the idea of a coop and run but lack space. Therefore an ark might be the answer. Even a large ark provides a manageable, lightweight, sturdy and a highly movable form of housing, an excellent starter pen for your Serama. They range in size, and different models cater for different needs, for example young birds would not require a nesting area, whereas an adult breeding pair would. They are relatively cheap to buy and if cared for, will last for many years. It is advisable to apply wood preservative annually, paying particular attention to the undersides of the ark.

This variety of caging is easy clean and perfect where space is limited.

A popular style of housing for breeders is the cage unit system. This allows several pairs or trios to be kept effectively in one large cage that in turn, minimises floor space. These units can be quite expensive but second-hand examples can be bought for a fraction of the cost. For the breeder, these units are invaluable as not only are they easy to clean and maintain, but breeding birds can be monitored more closely.

A common variation on these units is a converted outdoor shed that holds individual cages or stalls that run along the length of the structure. A backed table can be used and divided into sections that have a wire or caged front. The base can be scattered with wood shavings with a nest box in the corner. Additionally, a wire base can be fitted which would allow droppings to pass through and collect in a tray. This does have the benefit of keeping the birds' feet clean but is not an ideal method as it is very unnatural environment for them. It is vital for this kind of accommodation to be positioned inside a well-insulated outbuilding or similar, as the metal framework offers no warmth to the inhabitants.

Generally because of its overall size and weight, the units are fitted with wheels; ideal for moving outside for cleaning or simply to give the occupants some fresh air on warm days. If you would prefer to free-range your Serama, a form of rotation can be applied where different birds are let out of their cage each day. One disadvantage with this style of communal housing is that an airborne disease could affect the whole house if not quickly identified.

Whatever type of accommodation you decide upon, it is vital that your Serama bantams' welfare remains paramount.

This Serama cockerel weighs 405 grams. He has an excellent vertical wing carriage and a short back.

The breast could be improved as it should be full and carried high.

Choosing Your Serama Bantams

Before taking the exciting step of purchasing your Serama, it is important to think about what you expect from your birds.

Type, quality, colour, along with size, all are factors that require careful consideration. You also need too decide between the silkied or the straight-feathered Serama. Silkied Serama is so called because of its feather structure, which does not feature the barbs found on their straight-feathered 'cousins' and therefore the birds have a fluffy appearance.

When a Serama does not fully comply with the recognised standard, it is known as a 'pet quality' bird. These faults can be comb or wattle defects, over sizing or a bird with poor type. Consequently, these are unsuitable for breeding but they do make fantastic pets and are likely to be less expensive. These individual faults can be bred out of the birds by selecting the offspring with improved features and breeding these with similar enhanced birds.

A group of growers. Study how they react and whether they are active and alert.

If, however, you are keen to have breeding-quality Serama bantams then it is advisable to start with a pair, or two pairs of birds from a reputable breeder. A good breeder will be able to set you up with unrelated pairs. Buying two unrelated breeding pairs, enables the offspring from each pair to be bred together, furthering your breeding plan.

If you are unsure of what physical features you are looking for, try to compare the bird to the UK Serama Bantam Standard, which covered further on in the book. This will give you an idea of what to look out for, and – equally important – what to avoid.

Following the UK Standard, Serama bantams are no longer categorised into classes but instead their weight must be below the figure specified. However, it's likely that you will still hear breeders categorising their birds into classes, therefore you will need to know what the class sizes stand for.

The following figures show both the male and female class weights.

Mature Male Serama	Mature Female Serama
Class A – Up to 350 grams	Class A – Up to 325 grams
Class B – Up to 500 grams	Class B – Up to 425 grams
Class C – Up to 600 grams	Class C – Up to 500 grams

Breeders are aiming to produce small, dainty birds, therefore Serama of lighter weight are encouraged. A 280-gram Serama is very appealing but does make breeding more challenging due to the small size. It would be more advisable to start with a bird between 325-grams to 450-grams for breeding – these make the best reproducers.

Tiny hens tend to lay very few eggs, and those that are laid will be very small, thus causing incubation difficulties. Larger hens – in the region of 400-grams – lay better-sized eggs, making them more suitable for incubation, and the larger cockerels tend to be more virile. As Serama do not breed true to size or colour, it is possible to hatch birds of varying weights although usually the trend is that the greatest percentage will follow the size of the parent birds.

Serama at their heaviest weight, also make good breeders because of their larger size but ideally these should not be bred from as we are looking to create birds of a small stature.

As well as the physical appearance, it is vital when making your choice that the bird's whole demeanour and attitude is assessed.

Serama bantams are proud and confident little birds when reared correctly. Bullied by brood mates, poor rearing methods and dietary deficiencies can all account for meagre specimens and these are best avoided. Take time to observe how the birds interact; some can be quite rough and vicious with others. This does not only apply to males, as females can be quite merciless towards subordinate flock members. This is not always a bad trait as long as the bird in question is monitored when introduced to others.

The individual should not shy away or be flighty when approached and there should be no sign of aggression towards the handler. Birds that do exhibit these tendencies should never be allowed to breed as this is a trait that must not be promoted.

Serama that have been handled from a young age are well-socialised and will demonstrate complete trust towards humans. They should also be alert and active; birds seen to be hunched up and hiding in the corner are not good candidates, as this could be a sign of disease. Also have a look at how the birds are housed. They should be clean, well maintained and with little or no odour as you enter their environment. Signs of stale food and water, lack of fresh litter and a sour smell are likely indications of poor management.

Once you have decided on your chosen birds, you may box them and return home. If you have others, it is important that the newcomers are isolated and monitored for any health problems. They should be quarantined for approximately 10 days. You can now look forward to learning for yourself why this delightful little breed is increasing in popularity worldwide.

A pretty 318 gram Serama hen. She has an excellent high breast, full tail and her comb and wattles compliment her head.

Back length should be shorter but she is showing a good outline.

Feeding

There are many different types and varieties of feed on the market today and the choice can sometimes be a little overwhelming.
Serama bantams, like other chickens, are omnivores. Therefore their diet can consist of anything from seeds and grasses to insects and worms. Obviously if you free-range your birds, they will be able to exhibit their natural behaviour and forage for many of these things themselves. If, however, you decide to house your Serama then it is vital that their feed provides all the nutrients to keep them fit and healthy. Whichever feed you decide upon, it is important to check that your bird accepts and thrives on the diet. Try different varieties and find the one that is right for you and your Serama.

Chick Crumbs

As the name suggests, chick crumbs are fed to chicks from hatching through to approximately 10 weeks of age. They contain a high percentage of protein and optimum levels of vitamins and minerals essential for the first few months of life. The crumbs are designed for small beaks, with certain brands providing a smaller-sized crumb than others. Newly hatched Serama chicks can sometimes struggle with the size of the crumbs so these should be finely crushed using a blender, pestle and mortar or a rolling pin. Continue with the crushed chick crumbs until around 3 weeks of age or until they can manage the crumbs in their larger state.

Growers' Pellets

Growers' pellets are fed to youngsters from the age of approximately 10 to 16 weeks old. After this, your Serama can be treated as an adult and fed accordingly. These pellets contain the correct balance of energy and amino acids to maximise growth rate. Do be careful when purchasing your growers' pellets, look for a diet that is free from artificial chemicals and growth promoters. Again ensure that the pellets are not too large.

Layers' Pellets and Mash

Layers' pellets offer your bantams a complete feed that contains the correct balance of vitamins and minerals necessary for breeding adults. You can start feeding your Serama on these from the age of 16 weeks. This will prepare them for laying and breeding as it helps with shell quality and produces a good yolk colour.

Make sure that the pellet is not too big for your particular Serama otherwise it might have a problem eating it. If this is the case, you may find crushing or grinding the pellets makes for easier feeding. There are certain manufacturers that cater for the smaller breed of chicken and have pellets to suit.

A similar complete feed is Layers' Mash, sometimes called Layers' Meal. This is made from the same compound, but has not been compressed into a pellet. It can be fed dry or mixed with a little water. If adding water, do remove any leftovers as the soggy food soon turns stale.

Many breeders prefer mash as it takes the bird far longer to eat than the pellets and therefore they do not bulk up so easily. However there can be a lot of wastage as the bantams will scatter it in their search for the best bits!

Powdered mash has another disadvantage because is it can become stuck in the roof of the mouth, which causes the bird to gape and shake its head. If this happens, use a blunt instrument to remove the compacted food.

Mixed Corn and Plain Wheat

Mixed corn is a blend of wheat and split or whole maize. It can be purchased cleaned or, if bought directly from a grain merchant, may contain husks and wheatears.

It can be fed occasionally to your birds and is an ideal treat. A small handful scattered in the afternoon will provide enrichment by encouraging the birds to forage for their food. Mixed corn is a heating feed, perfect for the cold, winter nights but do not over nourish, as the maize is very fatty and irresistible to poultry!

You can also purchase whole wheat, which is less fatty. Plain wheat is ideal for feeding a broody hen, as she will not need to defecate as frequently as she would if fed on layers'.

Greens

Serama love to peck at vegetable matter. Cabbage leaves, sprouts, shredded carrots, boiled potatoes and other veggie household scraps form a welcome addition to their diet. Lettuce should really be avoided as feeding too much, can cause stomach upsets. Some Serama like the occasional fruit but always introduce this slowly and discard after a couple of hours. Never feed avocado, as it is poisonous to all birds. Gathering wild food is also appreciated but do be cautious as it may have been polluted by exhaust fumes or sprayed with pesticides.

Additional Treats

Apart from the supplements previously mentioned, some hobbyists find it beneficial to offer mealworms to their Serama. These can be obtained dried or live, and will be eagerly gobbled up!

These are perfect as the occasional treat and for encouraging nervous birds to trust their owners. The drawback with mealworms is that they are highly calorific, so a couple for each bird is ample.

Grit

Serama - as with other poultry - are unable to 'chew' their food. Instead they swallow it whole, and it then settles in the crop. This acts as a kind of storage tank. From here it travels to the glandular portion of the stomach where digestive juices are secreted to help break up the food. The food travels to the ventriculus or as it is most commonly known, the gizzard. The gizzard has a thick, muscular wall that uses the ingested grit to grind down the food. Therefore it is vital that we provide the Serama with grit in

Granite (left) and Crushed Oyster Shell.

order for them to process their feed and obtain the necessary nutrients. Granite or crushed oyster shell is perfect for creating the grinding process in the digestion tract. Granite comes in many different sizes and the smaller stones are ideal for Serama. Crushed oyster shell is more suited for larger fowl, ducks and geese, as the angular shape of the particles can damage the bantams' petite gullets. However oyster shell is an excellent source of calcium – vital for the laying hen – so purchasing the finer variety will help with the formation of good eggshells.

Adding a cuttlefish bone for your birds to peck at will not only supply calcium to their diet, it will also help to maintain a well-shaped beak.

Lack of sufficient calcium is often considered one of the potential causes for egg binding - the inability to pass an egg - potentially fatal to the hen.

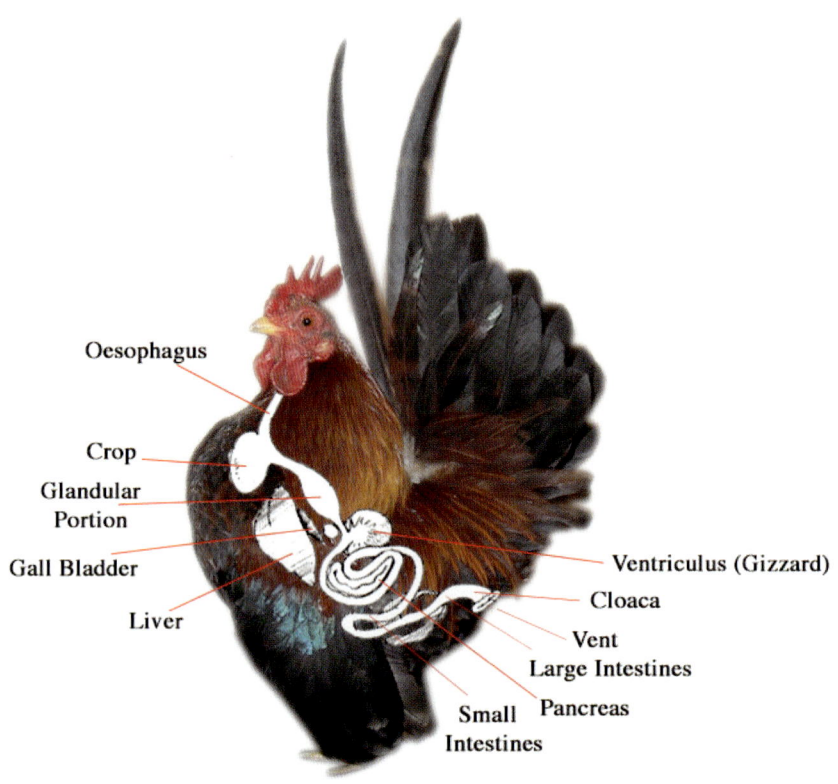

The digestive system.

Drinking Water

The diet of your Serama provides very little moisture therefore you must ensure they have access to clean water at all times. Their water can easily be soiled - droppings, bedding and food will all cause it to go stale, so it needs to be changed daily. By raising the water bowl off the ground, either by placing it on a brick or hanging from a chain, the water will remain cleaner for longer.

In addition, a simple clip-on bowl serves well for cages, keeping it off the floor and preventing spillage.

In the winter months, most owners will bring their Serama into warmer accommodation, so concerns that their water might freeze solid should not apply.

There are many different water containers and dispensers on the market, from plastic to galvanised metal. Plastic containers are easily cleaned, lightweight and inexpensive to buy but will eventually become brittle and will need to be replaced. On the other hand, galvanised metal will last a lifetime. However, if the outer coating gets damaged, rust can pollute the drinking water.

Whichever drinker you decide upon it is important that it is scrubbed clean with a brush occasionally to prevent the build up of algae – this is more likely to happen with water dispensers in a free-range environment.

A natural remedy for this unsightly problem is to pour a small amount of cider vinegar into the drinker, then wipe round and rinse. This keeps the apparatus clean and fresh and should be repeated monthly. Cider vinegar and its benefits are covered in depth later (see page 50).

Plastic water dispenser.

Breeding, Incubation and Rearing

There comes a time when simply 'keeping' these charismatic little birds is not enough. You are likely to find that you have become so passionate about individual Serama that you want them to live on. The only natural solution is to breed.
Breeding takes time and consideration since the main objective is to improve on the parent birds, therefore choosing the right pairing is essential.

Before you decide on your breeding pens, you need to study and absorb the standard for the UK Serama bantam. You will find a reference to this on page 42.
You are unlikely to have perfect birds, therefore assessing each specimen before pairing will help you match faults with non-faults that in turn, with time, will cancel out the defects.

Let's look into it in more detail.

Cockerel

Hen

The photographs above show two potential breeding birds. In order to see whether they are correctly matched, we need to consider a few points.

Do the birds possess any qualities that you would like to encourage?

Positives

   ✓ The male is showing great presence and arrogance as he stands in pose.
   ✓ He has a strong, vertical wing carriage of a good length.

- ✓ His tail carriage is high and vertical.
- ✓ He shows a pronounced chest.
- ✓ Short backed.

- ✓ The hen is showing a full and well-spread tail.
- ✓ Loose, profuse feathering.

- ✓ Both birds are of a small size (under 400 grams).

Negatives

- ➤ The male's feathering is rather 'tight' and not as profuse as one would expect.
- ➤ His tail should appear more 'full'.
- ➤ The hen is squatting and lacking confidence. **
- ➤ Her wings are not held vertical.

** Some hens do not possess the same confidence or the 'look at me' attitude as the males. Therefore they do not always look their best and this needs to be taken into consideration. The hen in the photo above is no exception. If she were standing confidently, her wings would be more vertical and her back would shorten.

After combining the pros and cons, you can see that these two Serama complement each other, because one makes up for the inadequacies of the other. Only a handful of the offspring produced will be an improvement on the parents but these birds can be grown on and used in your breeding pens.

Let's evaluate another pair – this time Silkied Serama.

Cockerel

Hen

Positives

- ✓ Again the male is showing great confidence as he stands in pose.
- ✓ He has a good, vertical wing carriage.
- ✓ Excellent high tail.
- ✓ Correct number of serrations on his comb.
- ✓ Relatively decent chest.
- ✓ Short back.

- ✓ The hen is short backed.
- ✓ She has a good chest.
- ✓ Excellent full tail.
- ✓ Nicely sized wattles.
- ✓ She has an acceptable wing length.

Negatives

- ➢ The cockerel's wings are far too short, causing the bird to look long in the leg.
- ➢ His comb and wattles are on the large side.
- ➢ Once again, the hen is squatting.
- ➢ The serrations on her comb are not clearly defined.

After examining both birds, you can see that the hen will complement the cockerel and vice versa. For example the small wattles on the hen will help to reduce the size of the wattles in the offspring, and the vertical wing carriage will encourage an improvement in the young.

Currently the Silkied Serama gene pool in the UK is very limited and consequently it is difficult to source unrelated birds. Breeding closely related birds is not recommended as we are looking to produce strong and healthy specimens. However in order for the conformation (type) of the Silkied Serama to improve, it may be necessary to use a straight feather Serama to achieve certain qualities.
The silkied gene is recessive, the dominant gene being the straight-feathered variety. So to achieve a 'new' silkied line, it would be necessary to use selective breeding for several generations.
This is explained with the following diagram showing what is produced if a Silkied Serama (recessive aa) is mated with a straight-feathered specimen (dominant AA).

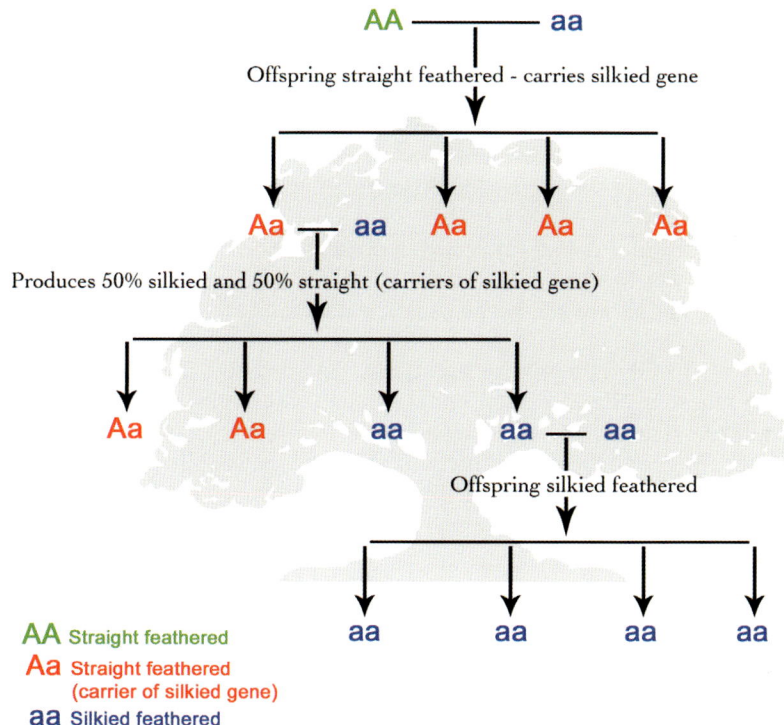

Along with assessing type, size must be taken into consideration. Serama do not currently breed true to size. This means that from a small bird, the offspring produced may range from the very small to the large, top-end weight. This is the case for all the classes/weights. However, it is preferable to breed from the lighter weight Serama in order to have a higher percentage of smaller specimens.

Do be mindful that small cockerels can have problems mating larger hens so are best kept with hens of a similar size. If he fails to fertilise the eggs, it could be an indication that the hen is too large.

The diet of the breeding Serama needs to be of a high quality and include the correct vitamins and minerals – vital for the laying hen.

As covered earlier in the book, it is most sensible to give your Serama a feed that is specifically intended for a 'productive' pair. Breeder pellets are available but these are more expensive than layers' pellets/meal and are not really necessary for Serama as they contain extra protein. Providing extra calcium and a mineral and vitamin supplement will help the pair remain in peak condition.

Artificial Incubation of Hatching Eggs

Once you have your breeding pair settled together, you can look forward to your first eggs. A Serama bantam pullet can be expected to start laying at anything between 16 to 24 weeks of age. A full, red comb will be an indication that this is imminent and you may see the cockerel treading her. Within a couple of weeks, you should have your first egg.

The eggs produced in a pullet's first laying season are not always viable – which means many of the incubated eggs may be 'clear' (unfertilised) or die during incubation. Eggs in her second season should make for better incubation.

Check daily for freshly laid eggs - then collect, date and note down which hen the egg is from. Store the eggs 'pointy' end down in a cool environment that is out of direct sunlight. Never place eggs intended for incubating in the refrigerator, as Serama eggs are very delicate and it could affect the hatchability. Incubate eggs within a week, as the membranes are stronger which gives you a better chance of them hatching successfully.

Suitable Incubators

There is a huge choice of incubators on the market today and they vary in cost. Some offer sophisticated temperature and humidity change warning alarms, while others boast auto-turn cradles and multiple-sized eggs trays. Aim to spend as much as you can afford, as you will benefit in the long run. 'Cheap' incubators sometimes fail to maintain the optimum temperature, and can cause humidity to fluctuate – to the detriment of the developing embryo.

Expect to pay between £100 and £200 for a good quality incubator. Within this price range, the equipment should include digital temperature readings, automatic turning and possibly a humidity reader – all helping towards a successful hatch.

Here the metal dividers roll the eggs from side to side.

If a broody hen were to sit on a clutch of eggs, she would turn them several times a day to ensure that that the embryo grows uniformly. If we are to artificially incubate eggs, we must replicate this behaviour.

Automatic incubators are so-called because they either sit in a turn cradle that gently tilts the eggs on its axis or the incubator uses a mechanism that steadily rolls the eggs in its tray. Semi-automatic versions incorporate a lever which, when pulled/pushed, rotates all the eggs in one movement. Manually operated incubators possess no turning facility and so the eggs must be turned approximately four times a day by hand – make sure you mark one side of the egg with an 'X' so that you know which eggs have been rotated.

The cradle rotates the incubator.

Next, you need to consider how many eggs you are hoping to incubate. If you only have a small number of birds, then it would be best to start with a small incubator. Running this equipment for just a couple of eggs is not really viable, but anything from 12-egg capacity upwards would be a good, beginner's incubator.

A still air incubator does not use a fan for ventilation; instead you may need to adjust the vent opening, depending on the humidity inside. These also tend to run at a slightly higher temperature than a fan-assisted, or forced air incubator.

Water trays in the incubator need to be filled in order to reach the correct humidity.

The humidity is checked by a hygrometer. This may be by using a dry bulb thermometer in conjunction with a wet bulb thermometer. The two readings are then compared and translated into the relative humidity. Alternatively a digital humidity reader is simpler to use and can be placed inside the incubator amongst the eggs.

An easy-to-read digital hygrometer.

Whichever incubator you decide upon, always refer to the instruction manual for that specific model as they can vary.

The Hatcher

To accompany your incubator you may feel it necessary to use an additional hatcher. This is a separate machine that is used, as the name suggests, when the chicks are about to hatch. The externally pipped eggs are transferred to the hatcher where the humidity is set to at least 55%. (Pipping is when the chick inside the egg starts to chip away at the shell membrane). The advantage of using a separate hatcher is that you are able to set eggs in the incubator at different times and when they are ready to hatch, you can place them in the hatcher without disrupting the remaining incubating eggs. Most incubators have the facility to be used as a hatcher but of course, the eggs inside must be at the same stage of development, as the humidity will be raised and the egg dividers will be removed.

Setting the Hatching Eggs

Once you have your incubator, you need to decide where it is to be positioned. A spare bedroom makes for an ideal environment as the temperature is unlikely to fluctuate dramatically and you will be able to keep a close eye on proceedings. Using an outbuilding with no heating or insulation will make it very difficult to maintain the ambient temperature. For the same reason, it is vital that you situate the incubator away from direct sunlight and radiators.

You will need to sanitise both your incubator and the hatching eggs to remove any harmful yeast spores, bacteria and fungi that could penetrate the eggs. A warm, moist incubator is a perfect breeding ground for these pathogens.

Place your disinfected incubator on a solid surface, such as a sturdy table, and switch it on.

- For hatching Serama bantams, the temperature inside the incubator should read **37.4 – 37.6°C**
- Humidity should be **38% to 45%** for the first 19 days and then increased to at least **55%** once externally pipped.

Do not add your eggs at this stage. The incubator should be run for at least 24 hours so that the temperature and humidity can stabilise. In the meantime, if using a manually-operated incubator mark each egg with a cross and remember to jot down the date you set the eggs for incubation. The 3-week incubation period will start from this date.

Once the incubator is ready, place the sterilised eggs between the dividers. The eggs should fit with just a small gap either side to allow for slight movement and will be positioned either on their side, or pointy end down. Generally the latter applies when you have limited space, as there will be room for extra eggs, but it is far more natural to lay the eggs horizontally.
Seal the incubator. You will notice that the temperature has fallen – do not alter the settings. The temperature will increase and within about an hour, the eggs will be in their optimum environment.
Inspect the incubator daily to make sure the water levels are sufficient.
To check whether the eggs are fertile, most poultry breeders would shine a bright light through the egg to show up evidence of blood vessels. This procedure is

called candling. Serama bantam embryos are very delicate and do not cope well with the intense light and temperature change. Therefore it is advisable not to candle your eggs during incubation. If, however, the 21-day period has lapsed and there is no sign of any chick, the egg may be candled to see whether it had been fertilised and if so, at what stage the embryo had died.

The Hatching Process

On average, the Serama bantam's incubation period is approximately 20 to 21 days. Pipping starts on the 18th or 19th day and once the chick has externally pipped, the eggs should not be turned again and the dividers should be removed. The humidity must be raised during this stage to prevent the membrane drying out which would in turn cause it to stick to the chick's down. If you are struggling to increase the humidity sufficiently, try adding a damp sponge into the incubator.

The chick inside has externally pipped the egg.

It is vital the incubator is only opened for a short period of time as the humidity level will drop and will take a while to correct itself. A hatching chick should take between 12 to 24 hours to fully emerge. If, after this period, the chick appears to be struggling, open the incubator and check that the membrane has not hardened. If it has, carefully dab a drop of warm water onto the hardened area to soften it and if possible, gently peel it back taking care not to break the blood vessels. Do not allow the chick to become chilled. Hopefully once the chick has regained its strength, it will be able to rid itself of the shell.

If you are hatching chicks from several different pairs, you'll need to ensure you can identify each chick. Simple removable plastic rings can be placed on their legs or if you are unable to find rings small enough, keep each bloodline separate until rings can be fitted.

The shell is slowly being chipped away from within.

The chick uses its legs to push open the shell.

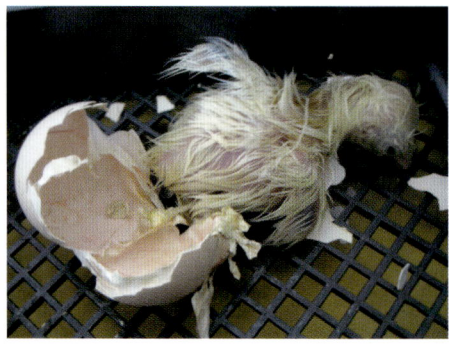

A dry, fluffed-up chick ready for the brooder.

The damp chick finally leaves the egg.

Introducing the Brooder

While the newly hatched chicks are drying out in the incubator, prepare your brooder. This is an area where the chicks are offered food and water while they are kept under constant heat. The heat source may be in the form of a heat lamp, an electric hen (a small table-like heater which the chicks sit underneath) or a ceramic heat emitter that does not produce any light. The temperature inside the brooder should be at least **35°C** for the first week of life, gradually reducing this by a couple of degrees every fortnight. There should be an area to where the chicks can retreat, away from the heat. It is vital that they are housed in a draught-free enclosure – preferably with solid sides, not wire. Draughts can cause problems for a chick in its later life or in severe cases, can result in its death while still young.

Once you have stabilised the temperature, you can add the chicks from the incubator. Newly hatched chicks are prone to 'splayed legs'. A specimen with this condition is unable to walk properly as its legs slide apart from one another. This problem can be avoided by covering the base of the brooder with a soft towel for the initial week or so of life as it will provide traction for the unsteady chicks.

For extra help, rubber matting can be placed on top of the towel. After about 2 weeks, the muscles in the legs have developed enough to prevent this from happening.

A brooder with solid, draught-proof sides and a ceramic heater.

You can clearly see a build up of droppings on this chick.

Another problem you may encounter is a build-up of droppings around the vent. Allowing the chicks to drink warmed water can help alleviate this but it is always wise to regularly check each bird. Leaving the vent blocked in this way can quickly cause the chick to become weak and appear bandy-legged, eventually the blockage will kill the youngster. Remove the cluster of droppings by gently pulling – in doing so, you may pluck away a small amount of down but this will assist in preventing the droppings from becoming stuck again.

The young chicks should be started on finely-ground chick crumbs and can be offered this 24 hours after hatching. Before this time, they rely on the absorbed yolk sac from the egg. Chicks will instinctively peck at the crumbs and can be further encouraged to do this by scattering a layer where they stand.

Warmed water should be placed in a shallow dish or use a saucer as Serama chicks are very small and liable to drown or become chilled if they get soaked. For added protection, place marbles or sterilised pebbles into the water bowl.

Notice the rubber, non-slip matting and shallow water dish.

If you have a poor hatch that has resulted in only one chick, it must be given some stimulation whilst in the brooder. Adding a small mirror and a soft toy will enable it to nestle amongst 'others' when you are not there. If possible, try to obtain another chick of a similar age as a companion – it doesn't have to be of the same bantam breed.

By the end of their first week you will notice the wing feathers coming through and the chicks will become more active and steady on their legs. At this stage you can exchange the towel for wood shavings or chips. Some breeders use paper as a floor covering, but this needs to be refreshed regularly to prevent their feet from becoming soiled. Sawdust is best avoided as the fine particles can cause havoc with the respiratory system.

Week by week you will see the chicks' feather up and exhibit sexual differences. Young cockerels will begin to 'spar' with one another to determine the pecking order amongst the group – young pullets are generally lower ranking and may be bullied. Keep an eye out to ensure that this fighting does not become too aggressive, as they may need to be segregated. If possible, it is advisable to separate the males from the females as constant pecking can affect their confidence. Check the head area for a change in appearance. In males, the comb and wattles will start to enlarge and redden – this can usually be seen at around 5 weeks, although certain strains, and the Silkied Serama, may take longer to develop.

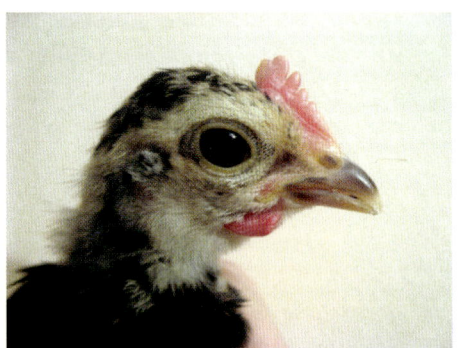

The reddened comb and wattles of the male.

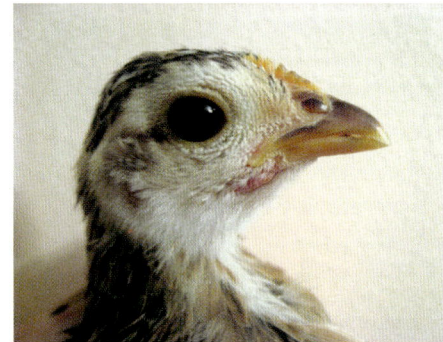

The pullet shows no, or very little, comb growth or colour.

From the age of 6 weeks, depending on size, you can begin to identify your growers with permanent leg rings. Not only will this help with record keeping but it will also tell prospective buyers the age of the birds. Leg rings can be plastic or metal, coloured or plain. You may like to have some produced with your initials printed on each ring.

The Poultry Club of Great Britain supplies plastic rings that are individually numbered and have the current year printed on them. Each year is a different colour so you can see at a glance when the bird was hatched.

If you do decide to buy leg rings from the PCGB then the size required for females is A10 and for males, A11.

If you are hoping to show your Serama, the PCGB leg rings are the only ones accepted on a show bird. All others must be removed before showing.

To place the ring round the leg, close the toes together making sure you include the back toe, then put the four toes into the ring and slide it upwards. If it is a little tight at the ankle, smear some Vaseline or olive oil into the area to lubricate it. As the cockerel matures spurs will develop on the back of his legs, and you'll have to lift the ring above the spur to prevent it from irritating the bird.

As the growers mature, you may need to house them in a larger enclosure so they can exercise their wings and legs. The chicks ought to be completely off heat now but they should still have protection from the cold, therefore house them in a relatively warm environment until fully feathered. Gradually the temperature can be reduced so that they will be happy living in a cooler area.

Never expect young Serama to be able to cope with the British winter. Although most adults can tolerate the low temperatures, the young Serama bantams are not yet fully acclimatised and lack the fat layer necessary to help them through colder spells.

This ring includes the bird number, breeder's initials and the current year.

You should be starting to see which chicks are showing potential; nice short back, vertical tail and wings, and prominent chest. Some strains take longer to mature and you may not see their promise until they are as old as 12 months. The Serama bantam is a slow-maturing breed and this can take up to 18 months.

At approximately 10-11 weeks, you can gradually reduce the chick crumbs and replace with growers' pellets. These should be fed to the youngsters until around 16 weeks of age, at which stage; they will be ready to progress to layers' pellets.

The Use of a Broody Hen

Serama bantam hens make perfect broodies. Not only do they save you a lot of time and money but also the chicks will be more acclimatised to cooler temperatures and yet still have the protection from the mother.

It is a good idea to use a hen that has gone through her first laying season. Sometimes young hens will not complete the 3-week incubation period and leave the eggs prematurely.

Most hens will lay approximately 8 to 12 eggs before becoming broody. To encourage this behaviour, replace her eggs with pot eggs. She will need a secure nesting area, lined with shavings, away from draughts and direct sunlight; a shoebox with one side cut away makes for an ideal nest box. She should also be treated with an anti-parasitic powder, as she will not be as meticulous in preening while incubating. To recognise a broody hen, look out for one that is reluctant to leave the nest and will fluff up and cluck when approached. Once she has reached this stage, leave her for 24 to 48 hours to ensure that she is settled with the nest and eggs. After this, you can replace the dummy eggs with the eggs you intend to hatch. As with artificial incubation, label the eggs so that you know which breeding birds they are from.

Many breeders ask whether the cockerel should be removed from a pen while the hen is sitting or brooding chicks. Most cockerels do not interfere with the sitting hen although make sure he is not sharing the nest at night, as it might expose some of the eggs to the cold. When the chicks hatch, he may either completely ignore them or help rear the newly hatched. If you opt to keep the cockerel with the hen, keep a close eye on proceedings until you are satisfied that he poses no threat. If you feel he is bothering the hen, remove him to another pen.

Many of the other bantam breeds make ideal broodies for Serama. Pekin, Old Dutch and Belgium bantams are all suitable surrogate parents but remember Serama chicks are very small so only use the light weight bantam breeds.

A Pekin bantam caring for Serama chicks.

The broody hen will leave the nest several times a day to feed, preen and to defecate. These large, smelly droppings should be removed from the pen to prevent her from transferring them to the eggs by her feet. After about half an hour she will return to the eggs to resume her duties.

After 3 weeks there should be signs of chicks. The broody will cluck to the chicks and wait patiently for them to all hatch. The chicks will start to emerge from the nest after 24 hours. You'll need to monitor this quite carefully to make sure the mother has accepted the newly hatched chicks. Sometimes for no apparent reason, the mother can attack the chicks, in which case they must be removed. A devoted mother will show the chicks to food and water, offer warmth and protection.

The broody hen will usually lose interest in the chicks when they are around 10 weeks of age, although this can take longer. She will not be so vocal and attentive towards the chicks and may even start to lay, in which case; you will know she is ready to leave.

The sequence of photos follows a day-old chick's development through to adulthood at approximately 30 weeks old.

Day-old chick still possessing the egg tooth on the beak.

Chick at 8 days old. The wing feathers are now in evidence.

The same chick at 5 weeks old. Currently showing no potential.

At 13 weeks old, the sex of the bird should be clear – here's a pullet.

The 16-week-old pullet is starting to show comb growth and reddening – these are signs that she should be coming into lay shortly.

At 30 weeks, this female is only just starting to show her type. You'll need to be patient because the Serama can take longer to mature than other breeds.

Exhibiting Your Serama

Being able to exhibit your Serama is one of the most thrilling experiences in poultry keeping and a real achievement. Not all birds are show birds, so you'll need to know what to look for in order to produce the best specimen. Type alone will not make your bird a winner as confidence really does shine through, producing a 'look at me' kind of arrogance.

It is always wise to join the breed club as this way you will be informed about forthcoming shows and events as well as receive newsletters and show results. It's also good to read about other fanciers' birds and experiences. The club will have details of the Serama bantams' Breed Standard. This is a document accredited by the Poultry Club of Great Britain that aims to maintain uniformity and to assist breeders of that breed's particular attribute. Currently in the UK, the Serama bantam is still being perfected but by having the Standard to refer to, it assures that the correct qualities are encouraged.

Table Training

In order to properly judge the Serama bantam at a show, it will usually be removed from its show pen and placed on a table. Not all judges carry out this procedure, but it's best to be prepared if they do.

You want your bird to show off its characteristic traits – high tail, vertical wing carriage and prominent chest. Some Serama will naturally exhibit a pose-like stance, while others will need to be trained to do so.

Ideally training should start when they are young, when the bird is beginning to show some confidence.

Stand the Serama on a piece of carpet or another non-slip material that has been placed on a table. The bird will probably squat the first few times it encounters the new 'alien' surface, it also takes time for it to feel confident when away from the others. Once it has become familiar with the carpet, it should start to stand in a more self-assured manner. At this stage, encourage the bird to display with a high tail, keeping its head as close to the tail as possible.

This is achieved by positioning the bird to show its side profile. With the bird looking to the left, place your right hand behind the cushion on the tail and hold it so that it is in the vertical stance. Now curve your hand around the chest and stroke underneath the wattles, this will cause the Serama to bring its head back.

Hold in this position until the bird stands still; wait a few seconds and then allow the bird to relax. The carpet is just one type of a platform that can be used with the performing bird, so gradually introduce other surfaces so that it becomes confident and able to pose whenever prompted.

This training takes patience but with time, the bird will associate standing on its own with posing and will exhibit this whenever it is asked to.

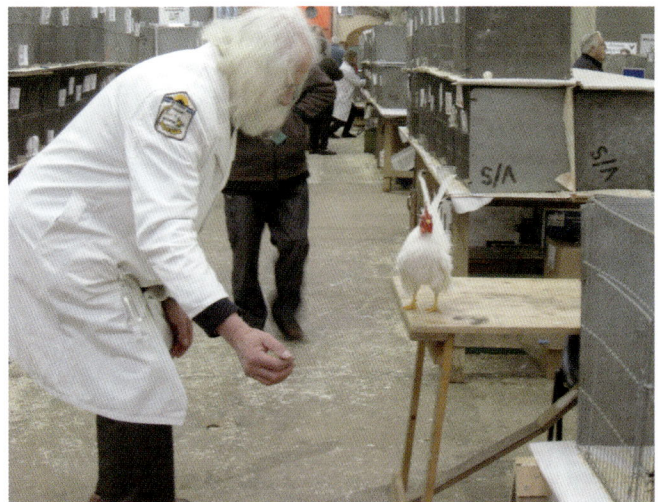

A judge assessing a Silkied Serama on the table.

Where to Start

Once you have decided that you have a suitable candidate to exhibit, start looking for a local poultry show to enter. As the Serama is still a relatively new breed, some of the smaller events may not have a specific class for it, so you'll probably have to opt for the Any Other Variety (AOV) classes, in which case you could be competing against other breeds.

It is recommended that you start with the smaller shows to build your confidence and understanding of the schedule and show rules. It's also a good idea to attend a few shows just as an observer so that you can see what other breeders are exhibiting.

Preparing Your Show Bird

Preparation is the key to success at a show. It is wise to have a bird or two on 'stand-by' in case the Serama you had planned to exhibit damages its feathers, goes broody or suddenly starts to moult – the process during which a bird sheds its feathers to make way for new growth. These factors would prevent you from entering your bird and so you'll want to have a replacement candidate lined up.

Feeding a good, balanced diet will help with feather and body condition. Adding a little Apple Cider Vinegar (no more than 25ml per litre) to their drinking water for a few months before a show can be very beneficial – it not only acts like an overall tonic but also improves feather strength and quality. Part the Serama's feathers to check for any external parasites and treat accordingly.

The bird should be brought into a caged environment about 2 weeks prior to the event, so that it can become accustomed to the penning cage used at the show. Ensure a thick layer of shavings is placed on the base and change this regularly to prevent a build-up of droppings.

Washing

About seven days before the show, the Serama will need to be washed. Doing this so far in advance allows the birds to preen, and thus replace the natural oils lost from bathing.

Items you will require:

- Three bowls, each large enough to stand the bird in
- Warm water
- Cup
- Dr Beckmann Glo-White for light-coloured birds
- Soap flakes
- Gentle Baby Shampoo
- Apple Cider Vinegar (ACV) for dark-coloured birds.
- Fabric softener for light-feathered birds.
- Toothbrush
- Small nailbrush
- Cocktail stick
- Nail clippers and file
- Moisturiser
- Towel
- Hairdryer

You may use some or all of the above, depending on the colour of the bird and its feather type.

Have the three bowls ready so that each can be filled in turn with warm water.

Your first objective is to clean the feet and legs. Shallow fill one of the bowls and stand the bird in it. Take the toothbrush and gently brush across the feet, not forgetting underneath. Work your way up the leg in a circular fashion. You may use a little non-perfumed gentle soap if you prefer for any stubborn areas.

Lift the bird up and using the cocktail stick, remove any remaining dirt from underneath the nails.

After the feet have been cleaned, check that the nails are not overgrown. If they are, use the nail clippers to trim any excess, being careful not to take too much off as this could result in bleeding. Shape the nail with the file for a more natural look. Discard the soiled water.

Next refill the bowl with clean, warm water to a depth that will come just above the bird's breast. Sprinkle a level tablespoon of the soap flakes into the water and stir them in. Once dissolved, hold the bird securely and lower into the water.

A hen being rinsed in ACV.

Gently part the feathers and dowse with water using the cup. The vent area and under the wings need to be soaked, as do the feathers around the head and neck, but take extra care around the eyes, ears and nostrils. Gently wipe the comb and wattles with a damp cloth.

Using the nailbrush, scrub the wings and tail feathers paying particular attention to the tips. ALWAYS brush in the direction of the feathers to prevent breakages. Again, if necessary, a gentle baby shampoo can be used to lift any remaining stains.

Check wing tips for any dirt.

If you are washing a white-coloured Serama, it is a good idea to use Glo-White in the water after the first wash as this will bring out the whiteness and remove any brassiness that may be showing on the feathers. Sunlight can cause this yellow staining on the feathers therefore white show birds may be better housed indoors. Only add approximately 1 to 2 teaspoons of the Glo-

White to the water – if you use too much it can result in a blue hue staining the plumage.
If you do not have Glo-White to hand, an alternative would be to use half a lemon squeezed into the water.
After this first wash, fill up the second bowl with clean warm water and rinse the bird thoroughly to get rid of the soap residue which can cause an irritation and may give the plumage a dull, greasy look. The nailbrush can be used again to ensure all traces are removed.

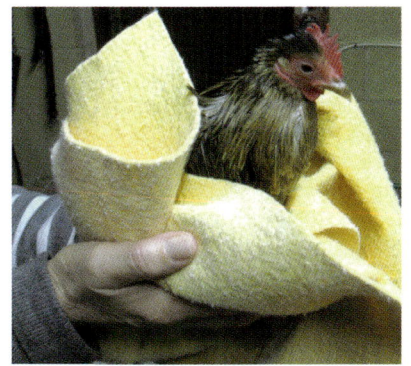

Fill the third bowl with warm water to a similar level and – depending on the colour of the Serama – add approximately 1-tablespoon of Apple Cider Vinegar, or 1 to 2 teaspoons of fabric softener.
The ACV will produce a healthy sheen to the feathers of darker birds and the fabric softener will improve the look of lighter specimens.

After this final wash, wrap the bird in a towel and remove the excess water. Transfer to a table and switch on the hair dryer, initially directing the air away from the bird. Keeping it on a medium heat (and monitor the temperature regularly), gradually introduce the warmed air to the bird. For straight-feathered Serama allow the dryer to blow in the direction of the feathers as this produces a smooth finish. However, you need to blow in the opposite direction on a Silkied Serama as this promotes a fluffy appearance.

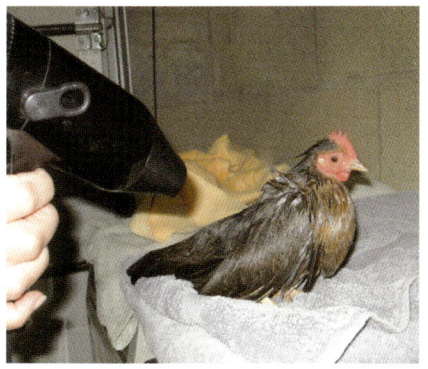

Once your Serama is fully dried, enhance their legs, feet, comb and wattles by applying a little moisturiser. Check that the beak is not overgrown. Trim and file any horn that may be over hanging – again, do not cut too much off.

The completed bird can now be placed back into its clean, warm pen to preen. Droppings will have to be removed several times a day to prevent any feathers becoming soiled.

Handle the bird as much as possible to get it accustomed to when it will be judged at the show and of course, practice placing your Serama in a pose on the table as an extra preparation.

The Day of the Show

The finishing touches can be applied on the morning of the show. It may be necessary to scrub the feet and legs again to clean off any droppings or staining. Once dry, apply a small amount of witch hazel gel to them, as this makes the legs look glossy and produces a good colour.

Ensure the nostrils and eyes are clean and rub a little Vaseline or olive oil onto the wattles and comb. This brightens up these areas and presents the bird as a picture of health. Your show bird is now ready to pen up.

A well-presented show bird takes second place.

One word of warning – there have been instances of birds being stolen, so you would be well advised to secure your show pens with a padlock once judging has finished. You will not be able to remove your birds until the appointed time. And sometimes you will be asked to leave the bird caged so that members of the public and breeders have time to view the different varieties of poultry.

It can be nerve-wracking – particularly when it is your first show – but the hard work and dedication will be worth it, and you can be justly proud of the bird you have produced.

Don't be discouraged if you are not placed early on, there are plenty of shows and as your knowledge increases, so will your success.

Returning Home

Once authorised, you can take your bird from the penning area and transfer it to your box ready for the return journey. Once you are home you'll need to put your bird into an isolation area.

Exhibitions are a breeding ground for a host of poultry-related illnesses and your Serama has been exposed to these risks. Stress will make the bird more inclined to pick up an infection. Therefore it is imperative that the bird is kept in the isolation room for at least a fortnight, by which time, any problems will have become apparent and can be treated accordingly.

Assuming everything is as it should be, the bird can be safely re-introduced to the rest of the flock. It's a good idea to add a mineral and vitamin supplement to your show bird's water as this will boost its immune system and help it to recover from its outing.

Three pairs of Serama are kept in isolation after a show for approximately 14 days.

The Standard for the UK Serama Bantam

As previously discussed, there is an official standard for the UK Serama bantam. This is affiliated to the Poultry Club of Great Britain and was established in order to perfect the breed by recording its ideal characteristics. This standard is shown below for your reference. All breeders should seek to achieve these qualities when breeding show birds.

Origin: Malaysia
Classification: True Bantam
Egg Colour: Varying from white to brown

General Characteristics:

Male

Carriage & Temperament: Assertive with confident bold stance yet calm and manageable. Should be easily handled and show no aggression. The bird should pose readily and when viewed from the side should create a vase-like or wide 'V' shape.

Type: Body well muscled with breast carried high, full and well forward. From above the shape is somewhat elliptical, tapering towards the tail. The back should be very short and covered by abundant hackle covering both the shoulders and secondaries and flowing onto the tail coverts giving the base of tail a full appearance. Tail should be carried high and upright at a 90-degree angle, parallel to the neck, and should be large and full. Main tail feathers should be long and broad and should overlap. The tail should be open and when viewed from behind should be open to an angle of 45 degrees creating an open 'V' shape. Sickles are slightly curved and protrude beyond the main tail. Side hangers and tail coverts should be broad, plentiful and well curved.

Wings: Fairly large in proportion to the body, they should be held in a vertical position just clearing the ground and leaving the feet partially visible. Shoulders should be set high on the bird. Primaries are long, of medium width, with secondaries moderately long and broad.

Head: Head to be small and carried well back. The single comb is small to medium in size with five serrations preferred, though more are allowed. It should be straight, smooth, free of folds or any deformities and tending towards flyaway type. Wattles are to complement the comb, smaller being preferred and free from folds and wrinkles.

Legs and feet: The legs are of medium length, straight and set wide apart to allow for full and muscular body. They should be strong and stable. Thighs should be of medium length and well muscled, with shanks of good thickness.

Plumage: All feathers should be in good condition with lustrous sheen. Body feathers to be full and profuse.

Colour: Male and female – the principal colours seen are:
White, black (with blue-green sheen), buff, red, partridge, wheaten, mottled, spangled and duckwing but any colour or combination of colours is acceptable and none to be penalised.
Comb, face and wattles bright red, though darker is acceptable in the darker colours. Eyes clear and bright with any colour being acceptable.
Any colour legs and feet are acceptable.

Female
The general characteristics are similar to those of the male, allowing for the natural sexual differences.

Weight:

Male	up to 600 grams (21.16 oz)
Female	up to 500 grams (17.64 oz)

Scale of Points:

Type and carriage	25
Temperament	15
Tail Carriage	20
Wings	10
Body	5
Legs	5
Feather Structure	5
Head/Comb/Wattles	10
Colour	5
Total	**100**

Serious Defects:

Lack of attitude, nervousness or shyness. Long back, low tail carriage, wry tail, cow hocks. Duck feet. Legs too short or too fine. Feathering on shanks or feet. Comb other than single. Any general defects. Weight exceeding the upper limit.

Silkie Serama: Silkie feathered Serama are acceptable. However, they should still display a high and upright tail, wide as in the smooth feathered. The comb should be moderate in size as in smooth feathered and not large as in the Japanese.

Problems and Ailments

However diligent your husbandry and care, you may at some point have to deal with a sick bird. It's vital that you familiarise yourself with what can go wrong, so that you can recognise the symptoms as soon as possible and treat the Serama in question.

Certain problems can be caused by poor management, lack of quarantine for new birds or by contamination from wild birds and vermin.

The following advice is intended as a guide to what to look for and how to treat the most common of problems.

SYMPTOMS	DIAGNOSIS	TREATMENT	PREVENTATIVE
Noisy breathing, mucous in nostrils and throat, causing bird to shake head. Watery eyes. Distinct, putrid smell from around the face.	Mycoplasma (Respiratory disease)	Isolate bird immediately and treat both the individual and its flock mates with antibiotics – Baytril can be sourced from your vet.	Keep wild birds away from your Serama. Quarantine new stock.
Yellowy/green diarrhoea. Lack of appetite. Breathing difficulties, usually with beak open. Depression and nervous disorders.	Newcastle Disease (notifiable)	No treatment – birds will have to be culled. Contact your vet for details on how to deal with such birds.	Newcastle Disease can be vaccinated against – recommended if you feel your birds are at risk. Spread through direct contact from infected birds.
Poor growth, lack of interest in food or water. Yellow, foamy, watery diarrhoea. Sudden death.	E-Coli (Colibacillosis)	Antibiotics plus the use of a probiotic may be beneficial if diagnosed early on.	Poor management is to blame for this problem. Can be caused by stale and out of date food. Adding a probiotic to the drinking water can reduce the E-Coli in gastrointestinal tracts.

SYMPTOMS	DIAGNOSIS	TREATMENT	PREVENTATIVE
Leg or wing paralysis. Pale skin. Young birds thin but with good appetite. Sudden death.	Mareks Disease	No treatment. Post mortem is often the only way to identify the disease.	Birds can be vaccinated as day olds and this is very effective. Never mix adults with youngsters as older birds can be carriers but not show signs of the disease.
Yellow matter in the ear causing the bird to scratch at it. Head shaking with it held at an angle.	Ear Infection	Soak some cotton wool in a salty solution; gently wipe the ear/s to remove any discharge. Antibiotics should be administered.	A dusty environment may cause this but nothing can be done to prevent it.
Hen or pullet appears to be straining. Looks fluffed up and in a 'huddled' position. Tail seems to look arched with excessive bobbing.	Egg Binding (egg bound)	Bring the bird into a warm room, such as an airing cupboard. The warmth will widen the egg canal making it easier to pass. If the bird is still struggling, try to feed it with some bread soaked in liquid paraffin. If no change, consult a vet.	Ensure the laying hens/pullets have access to grit and calcium. If you prefer, then a calcium supplement can be added to their drinking water.
No eggs. Broken eggshell. Yolk may be present on a bird's beak or chest.	Egg Eating	If the habit is caught early you may be able to stop it. Make a small hole in an egg and using a syringe add in some mustard, then place the egg back with the culprit. The spicy taste usually puts them off. Another remedy is to use a 'bit' as shown in the photo that prevents them from piercing the eggs.	Boredom, dietary deficiencies, lack of nest boxes are just a few of the reasons for egg eating. Ensure that eggs are collected daily.

SYMPTOMS	DIAGNOSIS	TREATMENT	PREVENTATIVE
Hen or pullet may be straining and on closer inspection, a red mass is seen protruding from the vent.	Prolapse	Using a salty water solution, apply liberally to the area to clean off any dirt or bedding material that may be stuck to it. Apply a little Vaseline to the mass and gently push it back into the vent. Usually this measure will suffice, but the problem may re-occur.	The mass is actually the oviduct that has been forced out when passing an egg. This problem can occur in pullets that have come into lay too soon.
Horn on the beak is too long which prevents the two halves fitting together. In bad cases, bird may feel underweight.	Overgrown Beak	Using small nail clippers cut back the horn, finishing off with a file to smooth the edges. Take care not to cut too much off or you could reach the nerve.	Free-range Serama do not tend to suffer from this problem as they naturally wear it down as they forage. Cuttlefish provides an excellent alternative for housed birds.
Toenails are curling round at an angle. May cause difficulties in walking.	Overgrown Toenails	Same as above, use nail clippers and file to re-shape. Check regularly as if left for long periods, the toe can become twisted and deformed.	Both free-range and housed birds will need their nails attending to as they are not running on hard ground. Giving them access to an area that is covered in aviary sandpaper may help.
Large mass of dirt and droppings encasing the toes.	Toe Balls	Soak the hardened balls in warm water to soften and then gently peel away. Care must be taken as pulling at the dirt may damage the toe.	Slack management is to blame – the problem occurs when Serama are left to walk around in dirty litter. Change ground covering regularly to prevent a build-up of droppings.

Internal and External Parasites

Parasites pose a real threat to your Serama. If your bird becomes unwell – perhaps with one of the problems already discussed – parasites can take hold and slowly weaken the bird further. Likewise, carrying a large number of parasites can lower the immune system making the Serama vulnerable to other risks.

Both internal and external parasites rely on their host for food. They survive by feeding on blood, feathers or utilising the bird's ingested food. If left untreated, they can cause considerable damage and even death.

SYMPTOMS	DIAGNOSIS	TREATMENT	PREVENTATIVE
The bird feels very thin and emaciated. Still eating well but remains light. White diarrhoea – late stage turns green. In heavy infestations you may be able to see worms in the droppings.	Worms (Roundworms and Tapeworms)	Flubenvet Intermediate Licensed Poultry Wormer. This is highly effective against roundworm, gapeworm, caecal worm, hairworm and gizzard worm. Dosage required is one level teaspoon per 4.5kg of feed. Add a little olive oil to the feed so that the powder sticks to it. MIX WELL. Feed for 7 days.	Clean and disinfect house regularly and do not allow a build-up of droppings. It is wise to worm your Serama every six months as a large quantity of worms can cause other problems to arise.
Blood-stained diarrhoea, listless birds. Poor body condition.	Coccidiosis	Treat the bird's water with Coxoid for 7 days.	This is the most common protozoan pest. Certain feeds contain coccidiostats that prevent the problem occurring, although most breeders would prefer to treat with a single dose rather than with a medicated feed.

SYMPTOMS	DIAGNOSIS	TREATMENT	PREVENTATIVE
Flakey, sore-looking legs. Scales lifted, possibly bleeding. Severe cases may make the bird lame and the ends of the toes may come away.	Scaley Leg	Benzyl Bensoate is most effective. Use a small paintbrush to apply generously into all the scales. Treat daily for 4 days then once 7 days later.	Regular cleaning and disinfecting the entire house will stop an infestation of this mite.
Bird appears irritated and will be scratching excessively. On closer inspection, lice can be seen amongst the feathers and clusters of eggs around the vent area.	Lice	There are various anti-parasitic powders available that will rid the bird of the infestation. Part the feathers and sprinkle the powder onto the body, being careful to treat from head to toe, including under the wings. Re-treat after 7 days.	Wild birds and their nests can spread lice. Whenever possible keep away from poultry to prevent transferral. Provide dust baths for your Serama and by adding an anti-parasitic powder to the bath, the birds will treat themselves as they bath.
Excessive preening and irritation. Tiny, dark specks can be seen at the base of the feathers – most prominent in white-feathered birds.	Northern Fowl Mite	Treat as above, and apply to the whole flock as these mites soon multiply.	Again wild birds are carriers. Ensure your birds have access to a dust bath. Monthly dusting with the powder, will keep mites at bay.

SYMPTOMS	DIAGNOSIS	TREATMENT	PREVENTATIVE
Anaemic-looking birds, weak, poor growth in young specimens. Free-range Serama not wanting to roost. Poor egg production and low/poor fertility. Tiny red dots can be seen in nest boxes, crevices and under perches.	Red Mite	You may have some success with Duramitex but in certain regions, the mites have built up immunity to the active ingredient. Poultry Shield is a multi-purpose cleaner that dissolves the waxy outer shell of the red mite – highly effective and non-toxic.	Even though red mites do not live on the bird, a heavy infestation can easily kill an adult Serama; therefore stringent cleaning methods must be applied. Disinfecting or the use of the Poultry Shield should prevent an infestation but remember to treat the whole house including crevices as these are perfect breeding areas.

The Use of Apple Cider Vinegar

Apple cider vinegar (ACV) is simply pure apple juice fermented into cider and converted to acetic acid. The resulting vinegar is a pale brown liquid that can be pasteurised or unpasteurised. It is preferred in the unpasteurised state, which will have a richer smell and contain dark sediment.

ACV has many health benefits and can be fed to chickens all year round. It can be used as a tonic to keep your Serama healthy and can improve egg production and fertility. Including apple cider vinegar in their diet can prevent or ease many of the ailments already mentioned.

Benefits include:
- ♦ Help prevents internal and external parasites.
- ♦ Helps support a healthy immune system.
- ♦ Improves feather condition – ideal for show birds to add lustre to the feathers.
- ♦ Improves shell quality.
- ♦ Reduces odour – the vinegar neutralises the ammonia smell.
- ♦ Increases vitality in chicks and growers.
- ♦ Can be used to treat minor wounds.
- ♦ Inhibits mould and algae growth in water and food containers.

The dosage rate is 25ml per litre of water fed for 10 - 14 days to establish healthy gut flora. After this time it can be used twice a week at the same dosage.
For chicks, growers and adult birds that are housed, ACV can be given all year round at a rate of 1ml per litre of water.

When? What? How? Why?

There are some questions that are asked quite regularly regarding the Serama, and many of them have been raised on the very popular British Serama Bantam Community forum that the author set up in 2006. Visit the forum at www.seramaforum.com

There's no such thing as a silly question; after all, we are still learning about this new breed.

Here are a few common queries and their answers taken from the forum:

Q: I keep several pairs of breeding Serama and one trio. One of the pairs is not very productive. The hen lays good-sized eggs but upon incubating, most (if not all) are clear. Do you think the cockerel could be infertile?
A: The problem may be caused by the size of the hen. If she is too large for the cockerel, then he will have difficulties in mating her. Try placing him with a smaller hen and see if fertility improves. It has also been recognised that certain birds are not compatible and just never produce viable eggs. The reason for this is unknown but this could be a factor explaining why your pair is unsuccessful. It is can help to clip some of the feathers from around the vent as excessive feathers, particularly Silkied, can decrease fertility. In the winter fertility can drop. Keeping your breeding birds in a warmer environment will not only increase activity, it will boost their wellbeing.

Q: I have successfully hatched three yellow chicks and one black. Is it possible to tell what colour they are likely to end up as adults?
A: That is one of the joys of Serama bantams – you never know how the chicks are going to end up. Their colour can change several times before they reach adulthood. It would not be uncommon for a yellow chick to turn out darker in its adult plumage. It really is a waiting game!

A once-yellow chick becoming darker as it matures.

Q: I have heard the terms – growers, pullets, POL, cock and cockerel used, but what do they mean?

A: The term 'grower' is used to represent a youngster between the age of about 9 to 16 weeks – as the name suggests, it is the time when the chicks do lot of their growing.

A 'pullet' is a young female, usually under a year old. After this it will be referred to as a hen. POL is an abbreviation of Point of Lay. This is a name given to a pullet after 17 weeks, up to the time she starts to lay. A cock is a male bird, older than one year; while a cockerel is less than a year old.

Q: Will Serama lay through the winter?

A: As with most poultry, a Serama hen's laying pattern will decrease as the days shorten. She may also start her annual moult. If, however, the bird is kept under light, it will trick her into laying throughout the winter. Never use lighting 24 hours a day as it can cause the over-strained hen to prolapse. Do allow the hen a period of rest.

Q: I am having problems with my hatching chicks. They all seem to be positioned the wrong way round in the egg. Do you know the cause?

A: It depends on how the eggs are being incubated. It is vital that the eggs are positioned either on their side or with the wider end up. Inadequate turning is another possibility, as is excessive turning at the time of hatching. Avoid incubating round shaped or particularly small eggs.

Q: One of my cockerels has a purple comb and wattles. He is fine in every other respect, but I am worried about what might have caused it.

A: A dark head is sometimes just a bird's natural colouration. However, if the colour has changed to this then it could be a heart or circulatory problem, in which case, there is nothing you can do. If he seems fit and healthy, there is probably no cause for concern.

This partially-clipped wing makes the hen exempt from a show.

Q: Annoyingly I have clipped the wing feathers of one of my hens and I am now thinking of showing her. Is there any point in entering her for competition?

A: Wing clipping in the show world is frowned upon. As part of being judged the wings will be spread to see the conformation therefore if this is missing, you will not be awarded points. You may even be asked to remove the bird.

Q: Can Serama fly well?
A: Straight-feathered Serama can fly very well but tend not to, preferring to walk. Silkied Serama can only flap a short distance because of their feather structure. Generally speaking, Serama are not a flighty breed, which makes them easy to handle and perfect pets for gentle children.

Q: I house my Serama in a stable over winter. Is this likely to be a warm enough environment or should I boost the temperature with an oil-filled heaters?
A: Always be sure that if you are using any sort of heat source it is safely positioned and not covered. An oil-filled heater sounds ideal, but it would be sensible to insulate your stable to save on electricity and prevent as much heat loss as possible.

A small, mobile heater. Perfect for preventing the temperature from dropping too low.

Q: I keep Serama and pigeons, and wondered if I could keep the two together.
A: Pigeons can be carriers of numerous diseases and parasites, which could be passed on to your Serama. Your pigeons may well be clear of any problems but if they mix with wild pigeons, the risk of transferral is greatly increased. It's best to keep them separate.

To have more of your questions answered, visit the British Serama Bantam Community forum at www.seramaforum.com

References

The following is a list of useful websites and books:

http://www.seramaforum.com
The British Serama Bantam Community Forum
A friendly and helpful forum dedicated to the Serama bantam in the UK.

http://www.seramaclubgb.co.uk
The Serama Club of Great Britain

http://www.poultryclub.org
The Poultry Club of Great Britain

Starting with Chickens: Katie Thear.
Broad Leys Publishing Ltd.
ISBN 0 906137 27 6

The Complete Encyclopedia of Chickens: Esther Verhoef & Aad Rijs.
Rebo Publishers.
ISBN 90 366 1592 5

Vinegar 1001 Practical Uses: Margaret Briggs.
Abbeydale Press
ISBN 1 86147 167 X